Moral and Ethical Issues in Human Genome Editing

A Statement of the CEC Bioethics Thematic Reference Group

cec

conference of european churches

"[God] has shown you, O mortal, what is good.
And what does the Lord require of you?
To act justly and to love mercy, and to walk
humbly with your God."

Micah 6:8

Moral and Ethical Issues in Human Genome Editing

A Statement of the CEC Bioethics Thematic Reference Group

Globethics.net CEC Flash 3

Edition CEC at Globethics.net
Director: Prof. Dr. Obiora Francis Ike. Executive Director of Globethics.net
Series Editor: Fr Heikki Huttunen, General Secretary of the Conference of European Churches CEC

Globethics.net CEC Flash 3
Moral and Ethical Issues in Human Genome Editing
A Statement of the CEC Bioethics Thematic Reference Group
ISBN 978-2-88931-294-8 (online version)
ISBN 978-2-88931-295-5 (print version)

Managing Editor: Ignace Haaz

Globethics.net
Head Office
150 route de Ferney
1211 Geneva 2, Switzerland
Website:
www.globethics.net/publications
Email: *publications@globethics.net*

Conference of European Churches (CEC)
Ecumenical Centre
Rue Joseph II, 174
1000 Brussels, Belgium
Website *www.cecEurope.org*
Email: ceg@cec-kek.be

All web links in this text have been verified as of April 2019

Table of Contents

2018, according to the task and mission defined in the Conference of European Churches roadmap document Forging our Future (2.2.6). Since CEC enjoys a special relationship as an observer of the Bioethics Committee (DH-BIO) of the Council of Europe (COE), it seeks in its bioethics work to be particularly attentive to the agenda and concerns of the COE.

Having in mind the vast scope of the issues covered by bioethics and biotechnology, the TRG on Bioethics proposed at the beginning of its work to concentrate on two major topics: new developments in assisted procreation and predictive medicine. These two areas of concern came together in the unique topic of gene editing. The major result of our work in this area is the text before you: Moral and ethical issues in human gene editing. It was submitted to the CEC Governing Board and Organisations in Partnership as a contribution to present discussions on these important scientific and technological advances.

Rev Dr Meego Remmel (Estonian Council of Churches) and, Dr Julija Vidovic (Serbian Orthodox Church).

PREFACE

Dr Julija Vidovic
Member, Thematic Reference Group
on Bioethics
Serbian Orthodox Church

We are pleased to offer to your kind attention this concise document, the fruit of an intense and dynamic collaboration, of the Thematic Reference Group (TRG) on Bioethics of the Conference of European Churches[1] which worked, from 2015 till

---

[1] Members of the TRG on Bioethics who actively participated in the development of the present document were: Dr Ulrik Becker Nissen (Evangelical Lutheran Church in Denmark); † Rev Dr Jean Boboc (Romanian Orthodox Church); Dr Konstantinos Kornarakis (Church of Greece); Dr Murdo Macdonald (Church of Scotland); Rev Dr Brendan McCarthy (Church of England);

This text addresses a significant recent "game-changer"—the emergence of CRISPR/Cas9 and other related technologies, which have the potential to allow the very precise modification of genetic sequences. This inevitably raises the question of whether – and how – these techniques should be used in humans. After presenting the potential therapeutic importance of genome editing, its implications for medicine with emphasis on the human germ-line modification, the document concentrated more specifically on the challenges it involves and the legislation which needs to be developed in accordance with the application of human gene editing.

Accordingly, even though the group recognised that many of the issues related to genome editing in plants or animals are important, comments and discussions in this paper focused specifically on human genome editing and concluded that "the notion that knowledge and the choices it offers might be our downfall is as old as the biblical tale of the Garden of Eden, but, in equal measure, history demonstrates the enormous benefits in health and happiness that come with responsible exercise of our intellect and powers of invention." Bearing that in mind, "the newest developments in genome editing will demand that we think again about how to balance hope and fear." While there are challenges uniquely associated with human genome editing, this topic was approached by employing an analytical framework which might be applicable to any potential bioethical innovation, examining issues of safety, efficacy, ethics and prudence.

This discussion document provided a fruitful basis for the organisation of an international conference on gene editing entitled "PLAYING GOD? The science, ethics and theology of gene-editing." This formed an essential part of the Churches' contribution to a timely and burning public debate. The conference was organised in Paris from 27 to 28 February 2018, with financial support of the Council on Interchurch Relations of the Evangelical Lutheran Church in Denmark, and hosted by the Protestant Institute of Theology and the Orthodox Institute of

Theology "Saint-Serge", with media partner for this event Orthodoxie.com.[2]

The aim of this conference was to stimulate an ecumenical and Europe-wide discussion on gene editing as Churches' react to current research developments, to take into consideration the variety of challenges and listen to Churches' experiences and reactions to these challenges, and to use outcomes of these discussions for future CEC work in bioethics. Importantly, we reflected together on the contribution of a Christian viewpoint in the bioethical debate on modern parenthood. This included consideration of societal and technological developments that are leading to new concepts and forms of parenthood, including the impact on cross-border practices and reproductive justice. In this sense, CEC remains the place par excellence for Christian churches to meet

---

[2] All presentations may be viewed at this link : https://orthodoxie.com/en/playing-god-the-science-theology-and-societal-issues-on-gene-editing-in-paris-27-february-2018/ and https://orthodoxie.com/en/playing-god-the-science-theology-and-societal-issues-on-gene-editing-in-paris-28-february-2018.

and reflect, allowing them to bring a common voice to our European society.

FOREWORD

Fr Heikki Huttunen
General Secretary, Conference of European Churches

The Conference of European Churches acts in the service of its constituency we engage current issues on the European ecclesial, social, and political scene. This publication reflects the most recent developments in bioethics and the challenging ethical discussion around them. Through a conference organized by the CEC Thematic Reference Group on Bioethics in 2016, and this publication, our purpose is to inform the Member Churches, National Councils of Churches, and the Organisations in Partnership of this conversation and to engage them in it.

The modification of the human genome is a good expression how the visions of science fiction and fantasising about the future are becoming reality. It is an important test case for our churches and our times whether ethical discernment has any role in this development, or if scientific progress has the prerogative to seek its course without any direction by other values. Gene editing is part of a broader context where cutting edge technologies present a multitude of new ethical and theological questions and challenges. Every new achievement opens new horizons, providing answers but also new questions.

Questions about genetic conditions and possible gene therapy will only grow as technologies advance and as researchers learn more about the human genome. In the case of the Down syndrome, for example, genetic testing during pregnancy for this particular condition is now very routine, with the number of babies born with the genetic condition declining as a result of abortion. Some countries have virtually eradicated it. This poses serious questions about the intrinsic value of human life, the dignity of each person, the definition of "illness" and "normality", the inclusion of different people in our communities, and the definition of "good life".

How do values and biases fit into the genes we choose to edit? What is the difference between treatment and enhancement? We need robust ethical tools to guard against implicit racism, ableism, misogyny, and so on, in our decisions. Sex selection of babies to be aborted is already a huge global problem—there is an estimated 126 to 160 million fewer girls in the world because of this.

Another serious challenge arises with the creation of "surplus" human embryos resulting in certain treatments. Should they be disposed of or stockpiled, or used for other treatments? The ethical and legal permissibility of experimenting on human embryos represents a clash with classical Christian views on life. Our theological and spiritual discussion needs to be holistic, taking into account also the consequences of these questions from the point of view of social justice and all humanity, including future generations. As noted in the text that follows "If human embryos were to become routinely viewed as commodities for experimentation or as a simple means to an end, this has the potential to have a detrimental effect on the ways in which human life is valued in society."

These type of challenges are part of ethical debates which need to continue, but at the same time they are becoming everyday possibilities in many

European countries. There is a conflict between the evermore rapid leaps in science and the requirements of ethical discernment.

Creativity is an essential aspect of what it means to be made in the image of God. It can contribute to an optimistic Christian view on science and technology. According to the biblical account of the creation God gave humans the task to toil and guard their environment, and God invited them to name the animals and birds. To work with Creation is thus an intrinsic human calling and not rebellion against God. However, if God's will disappears from this approach, sin and injustice take over. The way humanity treats the created environment as something to be sold and bought, altered and exploited, can equally be seen in the treatment of fellow humans as merchandise without faces and lives. A whole and favourable response to the Creator's call, therefore, requires an aspiration to follow God's will. This implies that ethical discernment, accounting for the consequences of 'progress', should give direction to scientific advancement. One could propose from the Christian point of view, that no limits need to be set to science, as long as interaction between the divine and eternal perspective and human responsibility is part of the picture. The churches have a valuable place in this discussion by recalling that human knowledge is always developing and after each answer there are more questions.

INTRODUCTION

Human genetic engineering. Long the subject of science fiction, genetic research has advanced dramatically in the last few decades, to the point where it has now become possible for us to attempt therapeutic genetic modification. A significant recent "game- changer" has been the emergence of CRISPR/ Cas-9 and other related technologies, which have the potential to allow the very precise modification of genetic sequences. This inevitably raises the question of whether- and how- these techniques should be used in humans.

Even though we recognise that many of the issues related to genome editing in plants or animals are important, comments and discussions in this paper will focus specifically on human genome editing. While acknowledging that there are specific ethical issues which may arise in considering applying these technologies in humans, it is important that human genome editing must not be treated in isolation as a unique topic, but is set within a wider framework of bioethics.

While there are challenges uniquely associated with human genome editing, these are not of such a distinctive nature that a general, principled approach to bioethics cannot be applied to this subject. Accordingly, the topic will be approached by employing an analytical framework which would apply to any potential bioethical innovation, examining issues of safety, efficacy, ethics and prudence.

However, before considering these specific issues, it seems important in the first place to present the potential therapeutic importance of genome editing, its implications to medicine with emphasis on the human germline modification, the challenges it involves and the legislation which needs to be developed in accordance with the application of human genome editing.

1

THE POTENTIAL THERAPEUTIC IMPORTANCE OF GENOME EDITING

In 1901, Sir Archibald Garrod identified alkaptonuria as the first known human genetic disease. Today, at least 4000 human diseases are recognised as being caused by mutations in single genes (monogenic diseases), and many more which involve the input of multiple genes (polygenic, or multifactorial) have been identified.

While many of the single-gene diseases are classified as 'rare', in total they may affect over 400 million people worldwide. Some, such as sickle cell disease, affect tens of millions of people around the world and are only 'rare' in certain parts of the world. For a tiny subset of these patients, allogeneic hematopoietic stem cell transplantation (allo-HSCT) or solid organ transplantation can be used to cure their genetic disease, but for the vast majority of patients there is no cure and at best they are treated by management of symptoms (for example, the intensive physiotherapy and other treatments required in cystic fibrosis, a disease in which the causative mutation has been identified).

The idea of therapeutic genome editing was born out of the realisation that the ideal therapy for monogenic diseases would be to develop a method which can correct the disease-causing mutations directly. However, as genome editing has developed in concert with continuing improvements in our understanding of the genetic contribution to non-monogenic diseases, the principle of genome editing is being considered not only to cure monogenic diseases, but also as an approach to more common diseases that have multifactorial origins.

The use of genome editing to cure monogenic disease is conceptually straight forward (genome editing being used to correct an underlying genomic typographical error or mutation), but the outworking of this is not as simple as many imagine. As an illustration of the complexity of what is being proposed in human genome editing, consider that there are approximately 3 billion (3,000,000,000) base pairs in the human genome. There are approximately 3 million (3,000,000) letters in the text of the Bible-so it would take almost 1,000 copies of the Bible to contain the same number of letters as there are base pairs in the human genome. To achieve the successful editing of a single base pair in the human genome, as is being proposed, would therefore be equivalent to changing a single letter in one of 1,000 copies of the Bible, and to do so at the sub-microscopic scale of DNA.

However, genome editing is not only a method which potentially provides a mechanism to modify single nucleotides. These techniques may be able to make more sophisticated and nuanced genomic changes, which could be used to cure more common diseases or to modify their course. Moreover, these changes could be transmitted to all future generations. As such, genome editing is potentially very powerful indeed.

The development of this powerful genome-editing toolbox, and the use of these tools to create a wide variety of genetically modified species, has brought forward the possibility of using genome editing in humans. This was highlighted by researchers in China, who used the approach to genetic modification in a type of non- viable early human embryo. [3] These experiments, though successful, highlighted the inefficiency and unpredictability of the process—findings that would have been predicted by animal experiments. These results clearly demonstrated that this form of genome-editing technology, even if judged ethically permissible or desirable, is not ready for human application. Nonetheless, these specific experiments and the general concept of genetic modification of human embryos have generated much discussion. Subsequently, scientists in both the UK and Sweden have been given permission to perform genome editing experiments on viable human embryos- though, crucially, not to allow these embryos to be implanted. Most other European jurisdictions have not yet allowed such techniques.

---

[3] Tripronuclear human zygotes, which are genetically incapable of developing but are nearly identical in concept to normal (diploid) zygotes

These experiments highlight an important aspect of the discussion: the status of the human embryo. In some forms of genome editing research, the 'participants' are early embryos ("pre-embryos" in some literature). While the use of embryos for research purposes is legal in some jurisdictions (e.g. up to 14 days post-fertilisation in the UK), this does not mean that the practice is unproblematic. While many may acknowledge that embryo research aimed at improving therapeutic outcomes for others can be acceptable in some circumstances, it is important that all human embryos ought to be treated with respect and their distinctive status recognised in keeping with the relevant Codes of Practice. (In the UK, the HFEA Code of Practice Principle 3: "[to] have proper respect for the special status of the embryo when conducting licensed activities.") It is becoming increasingly difficult to discern what 'proper respect' means in practice when large numbers of embryos (particularly in relation to those which have been created purely for research purposes) are used in experiments and then discarded. If human embryos were to become routinely viewed as commodities for experimentation or as a simple means to an end, this has the potential to have a detrimental effect on the ways in which human life is valued in society. The debate on genome editing presents another significant opportunity for society to review its understanding of the human embryo and to reaffirm ways in which the embryo can be accorded special status and treated with respect.[4]

It is not surprising therefore that there is much current interest in human genome editing, and some consequent pressure from researchers to move quickly in this area, given the potential benefits which some argue might accrue from future clinical applications of the technique. As with previous discussions on human fertilisation and embryology such as *in vitro* fertilisation (IVF), embryo research and mitochondrial donation, it is essential that this topic is thoroughly and carefully explored before any medical application and change in legislation are considered. Historically, all participants in these discussions have called for broad consideration of

---

[4] Genome Editing: Open Call for Evidence Submission by the Mission and Public Affairs Council of the Church of England, 2016.

the issues by representatives of many perspectives: scientific, philosophical and societal. This call was reiterated recently by members of both industry and research communities.[5] Consequently, many issues may require thorough examination as to how these topics might best be tackled, rather than indicating a fixed position.

---

[5] *Lanphier E, Urnov F, Haecker SE, Werner M, Smolenski J. Don't edit the human germ line.* Nature. 2015; 519: 410–1.

2

APPLICATIONS TO MEDICINE

Ever since the discovery of specific human disease genes, scientists have harboured hopes that the responsible mutations could be reversed with molecular approaches. In cases where a gene product is missing, it may be possible to provide a functional copy of the gene. While this has been successful in a few cases, several challenges have stood in the way of this approach, including delivery of the gene to the affected cells, safe and efficient integration into the genome, and immunological reaction to the therapeutic protein itself or the vector used for delivery,[6] or the activation of an oncogene caused by integration of the vector in its vicinity.[7]

Genome editing has several potential advantages over previous methods of gene delivery. In most cases, a mutant genetic sequence will be corrected or otherwise modified at its normal genomic locus, so problems arising from random integration are not an issue, and expression of the gene should be easier to control. In addition, the editing materials would be present only transiently in the cells, so only the edit itself would persist.

Some examples of genome editing applied to the clinic have already been published, including one involving the gene for CCR5, the required co-

---

[6] *Wilson JM. Lessons learned from the gene therapy trial for ornithine transcarbamylase deficiency.* Mol Genet Metab. 2009; 96: 151–7.

[7] *Hacein-Bey-Abina S, Von Kalle C, Schmidt M, McCormack MP, Wulffraat N, Leboulch P, et al. LMO2-associated clonal T cell proliferation in two patients after gene therapy for SCID-X1*. Science. 2003; 302: 415–9.

receptor in T cells for most strains of HIV-1.[8] Like this trial, the future therapeutic applications that are easiest to envisage are ones involving *ex vivo* treatment. The treated cells could be analyzed *in vitro* to ensure that the desired modification has been made, and successfully modified cells could potentially be enriched before implantation in the patient.

If therapies based on stem cells are developed, genome editing and autologous transplantation may also potentially be used there as well. In this scenario, cells which have been genetically modified in the lab would be transferred into the patient as part of a treatment. Direct delivery to tissues in the body, by contrast, continues to present serious challenges. For example, in the case of cystic fibrosis, where multiple tissues are affected, success would require delivery to epithelial cells deep in the lung.

The precision of genome editing and the potential to correct disease-causing mutations in the DNA

---

[8] *Tebas P, Stein D, Tang WW, Frank I, Wang SQ, Lee G, et al. Gene editing of CCR5 in autologous CD4 T cells of persons infected with HIV.* N Engl J Med. 2014; 370: 901–10.

sequence have always made the field conceptually appealing. The development of the CRISPR/Cas9 nucleases has meant that the pace of progress has increased exponentially. It is predicted that in the next decade many genome editing-based clinical trials will be developed by academics, biotechnology start-ups and pharmaceutical companies.

The exact nature of the therapeutic edit may depend on the interplay between the underlying genetics and the specific pathophysiology of the disease. Thus, one editing strategy might be appropriate for one disease but not applicable to another. A number of basic strategies of genome editing and tools are now available both to correct typographical errors and to make more sophisticated changes to the genome. The potential of genome editing to treat genetic, infectious, and acquired diseases is being considered, some of which may cause the engineered genetic change to be passed from one generation to the next.

3

HUMAN GERMLINE MODIFICATION

The methods for performing germline editing on nuclear DNA have been applied to other mammals, including primates,[9] and modification of non-viable

[9] *Liu H, Chen Y, Niu Y, Zhang K, Kang Y, Ge W, et al. TALEN-mediated gene mutagenesis in rhesus and cynomolgus monkeys.* Cell Stem Cell. 2014;14:323–8.; *Niu Y, Shen B, Cui Y, Chen Y, Wang J, Wang L, et al. Generation of gene-modified cynomolgus monkey via Cas9/RNA-mediated gene targeting in one-cell embryos.* Cell. 2014;156:836–43.; *Chen Y, Zheng Y, Kang Y, Yang W, Niu Y, Guo X, et al. Functional disruption of the dystrophin gene in rhesus monkey using CRISPR/Cas9.* Hum Mol Genet. 2015; 24: 3764–74

human embryos has also been published.[10] To achieve modifications of the germline which are passed on to subsequent generations, a number of approaches are possible, both performed in conjunction with *in vitro* fertilization and then gestation of the resulting embryo. One is to make the desired modifications in cultured cells and then transplant a nucleus from a successfully modified cell into an enucleated egg fertilized *in vitro*. This technique, called somatic cell nuclear transfer (SCNT) is problematic, because experience with several different animal species has shown that it is associated with a very high frequency of developmental defects, perhaps owing to the difficulty of reprogramming a somatic cell nucleus for all developmental functions.

The second approach is to deliver the editing materials (nuclease with or without donor DNA) directly into a fertilized egg, and to let the modifications take place there. In this approach, there is a significant chance that the resulting embryos would be

---

[10] *Liang P, Xu Y, Zhang X, Ding C, Huang R, Zhang Z, et al. CRISPR/Cas9-mediated gene editing in human tripronuclear zygotes.* Prot Cell. 2015; 6: 363–72.

mosaic for the modification, if some nuclease cutting occurs after cell division and the efficiency is less than 100 %. Although CRISP/ Cas-9 editing usually appears to very specific, there is also the possibility of off-target mutagenesis. It would be challenging to assess this at a sufficiently early stage.

A third approach to germline genome editing might be to proceed by modifying gametes before fertilization, then fusing these modified gametes in a form of IVF. This would require not only effective methods for delivering the reagents, but an understanding of the DNA repair capabilities of sperm and eggs.

4

(ETHICAL) CHALLENGES OF GERMLINE GENOME EDITING

The ethical issues around using genome editing that might result in the transmission of specific genotypes to future generations needs to be put into the context of activities already taking place that similarly affect the genotypic makeup of future generations. Two such examples are pre-implantation genetic diagnosis with selective zygote implantation and curing or helping patients with genetic diseases, such that they might not pass down their disease-causing mutation to their children.

The critiques focusing on the potential for editing the nuclear DNA of human gametes or embryos largely break down into two categories which are used in ethical analyses of many different kinds of technologies and human actions. The first might be called consequentialist; it focuses on the possibilities for improving the human condition, through the elimination of deleterious characteristics or mutations. It might allow people who carry such traits to have children to whom they are genetically related without the prospect of passing on problematic or dangerous conditions. To the extent that these changes would persist across the generations, it is argued that this could benefit not only the immediate offspring, but also all of the descendants of those who use the technology. Conversely, it is this same phenomenon - a change passed down through the generations - which increases concern about unintended effects, where disadvantages might outweigh any advantages that genome editing confers. Because these risks would be borne by those who had no say in the decision, it eliminates the most common justification for such actions- that the person taking the risk has made an informed and voluntary decision to encounter the risk. While this is true in every case of parental decision-making on behalf of a future or existing child, in those situations the rearing parents would share with the child both the risks and the possible benefits, thus adding some constraints on action. But when those risks and/ or possible benefits are largely felt by future generations, this constraint, in the form of self-interest and self-protection, is removed. Thus, is it legitimate for an individual to make a decision which affects all their future descendants?

Due to the intrinsic uncertainty about downstream effects, it is necessary to consider the precautionary principle, which demands a justification before permitting any risk-creating activity, with risk being defined both in terms of known hazards and unknown possibilities. The latter is, by definition, incapable of measurement, leading to the criticism that the precautionary principle can be stretched into a generalized prohibition. In cases of known risks- such as devastating genetic diseases- some might argue that any risks associated with genome editing procedures are acceptable. To NOT use a God- given ability would, by this argument, be wrong. At the same time, it must be admitted that it is impossible to confidently predict ALL of the consequences of genome editing- whether of introducing deleterious traits, or by losing unanticipated benefits to retaining particular alleles.[11]

As to the justification for taking risks, a variety of means already exist to avoid passing on problematic traits, including the choice to forego biological reproduction, the use of donated gametes and embryos, or the use of pre-implantation and prenatal diagnostic techniques to avoid the birth of an affected child. Even while acknowledging that the option of embryo selection or selective abortion will be unacceptable for many, the availability of these alternatives may be seen as a means to diminish the prospective benefits of genome editing, by measuring those benefits solely in terms of marginal increases in personal choices and good birth outcomes. However, an important consideration in germ- line genome editing is that this reasoning only applies to the persons making the initial choice to edit the genetic sequences: all subsequent generations must

---

[11] For example, heterozygote advantage in the case of sickle cell hemoglobin, where a mutation confers advantage in resisting malaria infection.

bear the consequences (good or bad) of that choice, while clearly having no say in making the decision.

Another thread in consequentialist argumentation concerns the wisdom of any effort to alter the human condition through genetic manipulation. Even before the elucidation of the mechanism of genetic inheritance, societies across the world had eras in which they viewed selective breeding as a means to ensure the superiority of any resulting children. With the publication of Darwin's works, and their manipulation into social theory by Herbert Spencer, a new age of 'scientific' eugenics was born. Couched in terms of social hygiene, it attracted followers from all parts of the political spectrum and combined crude understandings of genetics with a host of cultural prejudices. Not surprisingly, it led to ugly decades of the worst form of eugenics, with mass involuntary sterilisations and mass murder. [12] Genome editing, like its less efficient predecessors (including choice of gamete donors, or pre-implantation selection of embryos), is seen by some as an opportunity

---

[12] *Epstein CJ. Is modern genetics the new eugenics?* Genet Med. 2003; 5: 469–75.

to clear deleterious traits from the family line, while being criticized by others for its echoes of simplistic and cruel notions of genetic superiority and inferiority.[13]

A second standard form of ethics analysis focuses less on specific consequences and more on some set of fundamental principles of right and wrong, considering the appropriate scope of human control over the planet and the species which inhabit it.

Many see the act of choosing to intervene to edit the genome as a usurpation of God's role in mankind's existence. While some see choosing as 'playing God', others see it as 'playing human'- exercising an ability given to us by God, as God's agents, part of the stewardship imputed to us. In this view, being such a partner means taking an active role, and 'artificiality', far from being wrong or evil, is rather a sign of humanity's constructive contribution, a sign of doing the duty God has given. Furthermore, it is argued, humans can actively engage in furthering the overall state of humanity by intervening in the works

---

[13] *Center for Genetics and Society. About human germline gene editing. 2015.*

of nature, when the goal is to achieve a natural good, such as health or fertility.

Nonetheless, there remain important issues to be resolved. These include developing a regulatory framework which is tailored to the underlying technology rather than one that is based on a different therapeutic foundation (such as small molecules or antibody biologics). There is also a need to develop safe and effective mechanisms to deliver the genome-editing machinery to a wide variety of tissues *in vivo*, including the liver, eye, muscle, heart, and brain. Finally, a flexible and adaptive regulatory framework needs to be developed to take into account the ethical and scientific issues around the potential use of genome editing that might alter the genetics of future generations ('altering heredity'). This framework needs to take into account the diverse group of stakeholders who are affected by the issue and must respect culturally different perspectives.

5

LEGISLATION AROUND HUMAN GENOME EDITING

In different countries, including European countries, legislation around experimentation on human embryos and about the appropriate degree of human control over its environment and its destiny, have been shaped by different histories and religious traditions.[14] It should therefore be no surprise that the use of genome editing is likely to be allowed to proceed at completely different rates among countries, cultures and regulatory systems. To address

[14] Engelhardt Jr HT. Global bioethics, theology, and human genetic engineering: the challenge of refashioning human nature in the face of moral and religious pluralism. In: Pfleiderer G, Brahier G, Lindpaintner K, editors. GenEthics and Religion. Basel: Karger; 2010. p. 40–51.

this reality, a group of scientists, lawyers and ethicists met in early 2015 in California; their discussions led to a call for a moratorium on human applications of germ-line editing. This was quickly followed by announcements by the US National Academy of Sciences and National Academy of Medicine, and by the Hinxton Group (a self-organized international group of scientists and ethicists), which issued statements on genome editing technologies and human germline modifications. While there is consensus among the various groups that technical advances are necessary before human germline applications should be undertaken, there appears to be a tacit assumption that such manipulations will ultimately go forward. In this context, various recommendations have been made that research on genome editing in human embryos should proceed under strict guidelines. While acknowledging the ethical concerns, many caution against over-regulation, which could inhibit orderly progression towards legitimate uses of the technology. In the UK, the Human Embryology and Fertilization Authority (HFEA), has authorised a number of specific proposals for the use of genome editing on human embryos, including one to investigate the causes for repeated miscarriages.

6

SAFETY

While safety is often viewed as a matter for scientists and clinicians to determine, issues of safety are far-reaching and require the involvement of others outside research and medical communities. While it is clear that research (and any subsequent clinical applications) ought not to place human participants at notable physical or mental risk without prior informed consent, it is less clear where the line is to be drawn for 'notable risk', who or what might be considered as 'human participants'. It is also important that this discussion includes social as well as physical and mental aspects of safety.

Few intrusive clinical procedures, conducted either for research or directly for therapeutic purposes, are entirely risk-free. Procedures necessary for obtaining ova for research or treatment, for example, carry an element of risk to donors, but these are deemed by the relevant regulatory authorities and others to be within acceptable limits. Similarly, the level of risk for IVF procedures is now well documented and closely monitored with participants enabled to give informed consent. Conversely, the necessary benchmarks for mitochondrial donation have not yet been reached. It is essential that margins of risk are clearly established for all participants in the clinical application of genome editing research, and that no procedures are authorised until it can be demonstrated that they fall within acceptable risk levels. Parliaments need to be assured that such is the case.

Ensuring that the offspring of any 'genome-edited' children will not suffer any detrimental effects is a major safety concern. By definition, it is not possible to know how subsequent generations of children might be affected until they exist. Animal experimentation cannot provide a sure guide in all instances. This has the potential to become a significant obstacle in introducing genome editing for clinical purposes unless convincing evidence can be produced that any associated risks have been minimised. Equally, a watertight guarantee would have to be put in place that any emerging problems would halt the continued use of the technique until they are fully resolved. However, as has been seen with some aspects of IVF, it may be decades before some problems are identified.

If genome editing were eventually to result in clinical procedures being authorised, the need to protect all participants from unacceptable risk is paramount. This must include donors, prospective parents and, primarily, prospective children. The emotional pull to seek to ensure that children are born without debilitating or life-limiting conditions must not outweigh the need to ensure that emerging techniques are used only if the risk of introducing other debilitating conditions to children and their offspring has been effectively precluded.

The social risks associated with introducing clinical treatment that might be available only to some prospective parents (particularly if treatment is to become primarily available through the private sector, as is currently the case with IVF), must not be minimised. The potential for increasing social disparity is a real one that needs to be realistically analysed and satisfactorily addressed prior to any change in clinical practice being introduced.

Similarly, the potential impact of genome editing on societal attitudes to disability requires careful consideration. While it would be wrong to suggest that minimising or eradicating some causes of physical impairment ought not to be pursued because of such concerns, the issue deserves to be thoughtfully examined.

Nobel Laureate Bob Dylan said: "Money doesn't just talk - money swears". To what extent will commercial interests be a catalyst for future developments or a pressure to drive it faster and further than is appropriate? What are the political choices involved in pursuing certain lines of genetic modification and not others? The possibilities of genetic "haves" and "have-nots" would be a very real risk. In his novel "Brave New World", Aldous Huxley explored the idea of applying genetic engineering beyond the area of corrective or preventative medicine into changing human characteristics - the notion of "improving" the human being, and its political corollaries, such as engineering a "master race" and a "slave race".

The borders between treating illness and enhancing human health are porous as are the borders between enhancing health and enhancing human performance/attractiveness. Vaccination, for example does not treat illness, but enhances the human body to resist infection. Similarly, nutrition does not only have an effect on health, but also on physical and mental performance. If genome editing were to become a clinical reality, the implications for human enhancement would need to be fully explored alongside the potential for furthering social disparity noted above. It would be an extremely difficult task to know where to draw the line along the continuum from treatment of illness to performance enhancement, but lines would have to be drawn to identify and to avoid unacceptable uses of genome-editing techniques.

Although genome editing approaches have been used in a variety of organisms, their use in humans is currently very much less well explored. However, some of the claims being made for the potential of genome editing in humans may be over-optimistic. Biological systems are not machines, so just because a technique works in a one species does not mean that it can easily be used in other species. For example, although Dolly the sheep became the first artificially cloned mammal in 1996, it was only in 2015 that dogs (another mammal) were successfully cloned using the somatic cell nuclear transfer (SCNT) techniques which were deployed in creating Dolly.

7

EFFICACY

It is essential that, from the outset, each application for research and any subsequent clinical procedures are absolutely clear in what they might realistically achieve. Equally, public discussion of the potential for genome editing to address illness must be measured. Potential clinical applications of genome editing must be explained and discussed in detail rather than promoted with generic claims 'to wipe out' certain medical conditions. (To achieve this aim, every potential parent in the world would have to undergo genetic profiling and every potential mother would have to agree to IVF; clearly an unrealistic prospect. Even this would not prevent the re-emergence of a condition due to novel mutation). Prospective parents must not be given false hope, and raising expectations which are not likely to be met for many years should be avoided.

Germline genome editing will neither significantly affect the numbers of healthy children to be born, nor will it affect the health of existing children; these realities need to be widely understood. Preimplantation Genetic Diagnosis (PGD) already provides a well-tested means by which clinicians may select 'healthy' embryos for IVF implantation while gamete donation from a third party enables potential parents, diagnosed as 'high-risk', to give birth to healthy children. There are, however, circumstances in which PGD is not effective and genome editing could enable some prospective parents to give birth to their genetically related children where a defective genetic sequence which would otherwise be passed to all of their offspring has been repaired.

While it is generally assumed that the ability to edit genetic sequences would have a profound effect on human health, this may not necessarily be the case. Biological systems are replete with redundancy, so that if one system fails, another pathway may be able to partially compensate for this.

A relatively small number of human diseases are attributable to a defect in a single gene. Although our understanding of the human genome may have increased, the vast majority of both diseases and normal human characteristics - height, physical endurance, artistic ability or intelligence - are highly complex "multi-factoral" phenomena. They involve the interactions of many different genes and any number of environmental influences and individual choices. Even supposedly simple things such as the inheritance of hair and eye colour seem to be complex. This is even before taking into account epigenetic and conformational aspects of the regulation of gene expression.

Many medical conditions result from the interplay between the products of numbers of genes. The mechanisms underlying these processes are often poorly understood and may not be able to be remedied by current genome editing techniques. Additionally there may be interactions between genes which do not come to light for several generations and are only observed through population studies. At the very least, much greater research into the interrelatedness of genes is required before the clinical application of genome editing is likely to prove efficacious (or safe) in all but a few cases.

To talk of "improving the human race" by means of genetic editing may seem to be science fiction, quite apart from any ethical objections. In addition, it would unfortunately be impossible to discount the potential of genetic abuse if ever the technology were feasible. Already, genome editing has been used to modify animal species for what might be termed frivolous reasons (e.g. making micropigs as pets, or fluorescent rabbits). While nobody is currently suggesting that humans be modified in similarly frivolous ways, if human genome editing is to be postulated, it is necessary to also ask what constitutes an improvement, how any improvement would be recognised as such, and who would have the authority to decide and control it?

8

ETHICS

Issues of safety and efficacy have ethical implications, but there are additional ethical concerns which need to be explored in a discussion of genome editing, before changes in practice or legislation ought to take place.

A number of principled questions can be applied to bioethical issues: what course of action might best be seen as life-affirming? Which best cares for vulnerable people? Which best contributes to building a cohesive and caring society? Which best respects individual freedom?

Affirming life requires an examination of the nature and value of human life and the limits that ought to be set for intrusion into both the life process and individual human lives. While it is customary for many to value life in instrumental terms or in terms of the value an individual might set on his or her own life, many would argue that life ought to be viewed in a wider context and that human life has an intrinsic value apart from its instrumental use or the value set on it by individuals themselves.

Many would argue that the human being is a complex unity, a whole entity more than the sum of such subdivisions as "body, mind and spirit". Human beings are "dust" in the sense of being made of the same physical stuff as the rest of creation. Yet they are more than animated dust- they have the breath of life. Even more than just living or existing, human beings can be described in terms relationship - to ourselves in self- awareness, to other human beings in family and community, and to the rest of creation. Each of these relationships leads to responsibilities, which sets bounds to what we may do within such a relationship- bounds which may be applied to genetic technologies.

This means that we ought to view life both as set within the whole environment of human relationships and as part of a continuum from one generation to another. Humans move along a spectrum which begins biologically with human genetic material arising from their parents (gametes) becoming separate, though dependent, human life (zygotes). Individuated human life emerges after the blastocyst stage and individual human persons emerge, although there is no agreement as to when this final stage is first reached. What value a human life is to be given cannot be determined by taking a simple 'snapshot' of where that life is on the spectrum at a given moment. Both potential and history ought to be taken into account: a much sought child at the embryonic stage is more than 'a collection of cells' to its parents, just as a much loved relative in end-stage dementia is much more than 'a shell' to his or her family. At the very least, human embryos ought not simply to be seen as routine means towards an end, created purely for the purposes of research; we need to reflect on the ethics of bringing human life into existence in order to experiment on it and then end it. While a positive case can be made for using for research 'spare' embryos (embryos originally intended for reproduction but which, for various reasons, could not be implanted), great care must be taken over the impact that genome editing research might have on the value that we place on human life.

The long-term effect on how society views the early human embryo must be seen as part of the cost-benefit analysis in assessing the usefulness of genome editing technologies.

Currently, experiments on living foetuses are not permitted, but in order to test the viability and efficacy of genetic changes made to embryos there is a real possibility of pressure to allow some of these to develop to the foetal stage, before licensing the technique for reproductive purposes. The desire by some researchers to extend the moratorium on research beyond current limits is a cause for concern. The present limit recognises the importance of the emergence of the 'primitive streak' heralding the beginning of neural development; regardless of one's views on the status of the embryo and foetus, it is reasonable to argue that as development proceeds along the human life continuum, increasing care ought to be taken with regard to how it is treated. If the beginnings of neural development do not provide a watershed for research, it is difficult to see where else the limit might be placed.

Currently, the possibilities opened up by genome editing are highly regulated in most western countries. That may not always be the case in other jurisdictions, and once the "genome editing genie" was out of the bottle, it would be extremely difficult to put it back in. This is a very fast-moving field, with which regulation inevitably struggles to keep up as application of the technology races ahead. In a parallel situation, the UK revised its regulations on what is permissible in terms of experimentation on human embryos less than a decade ago: the moratorium placed in that 2008 legislation on the gestation of embryos which had been experimented on beyond 14 days after fertilisation is now being challenged, and would of necessity be removed if human germline genome editing in a clinical setting were allowed.

To argue that human life has an intrinsic value apart from its instrumental use or the value placed on it by individuals is not, in itself, to determine what that value is, but it is to assert that there is something unique and special in being human. Christian theology grounds this in the concept of the Image of God, but it is not necessary to employ this religious understanding of human life to agree that human life is, in some respects at least, distinctive from other forms of life. That human life has an intrinsic value that ought to be respected is an innate understanding of many. It underlines much of human rights and criminal law as well as health and social care and can only be abandoned at our peril.

Caring for the vulnerable covers a wide variety of interests including prospective donors, parents and children. While it might be contentious to claim that embryos are vulnerable (and if they are, they are not vulnerable in quite the same way as adults or children), the possibility of extending the current limits for research brings into focus the ethics of deliberately creating, utilising and truncating human life even if these early biological lives are not to be equated with human persons.

Caring for the vulnerable must also be reflected in the ways in which those who donate gametes or embryos for research are given full and accessible information with regard to the use to which their donations will be put. It also requires that the recipients of any clinical treatment be made fully aware of associated risks and limitations as well as of the means whereby their treatment became possible through the use of embryo research. For many people this might not present an ethical problem, but for some it will and they ought to be given this information as part of the consent process

Paramount, of course, is the welfare of any children (and their offspring) who might be born, following genome editing. This welfare extends to psychological and social aspects of their wellbeing. Tremendous care needs to be taken to ensure that every aspect of their health and wellbeing is explored before any treatment ought to be authorised.

Building a cohesive and caring society is close to the heart of every civilised state and many advocates of genome editing view the technique as having the potential to contribute towards achieving this goal. At the same time, it is essential that all aspects of the potential implications of genome editing for society are fully explored. The 'law of unintended consequences' has an inexorable habit of exercising its influence. As previously stated, issues of social disparity, attitudes towards disability and the boundaries between treatment, health and performance enhancement require rigorous examination. Allocation of resources to genome editing over against other techniques that address some of the same issues such as stem cell research needs also to be carefully weighed.

Respecting individual choice is an important aspect of health care. For some, however, it has become the over-riding principle which all others serve. However, while respecting individual autonomy, the exercise of such autonomy must be principled and it is necessary that the context for principled autonomy is set by the affirmation of life, care of the vulnerable and the creation of a cohesive and caring society. Clearly, this places limitations on the freedom of individuals and the amount of resources that ought to be utilized in enabling them to pursue their aspirations within a framework that promotes the common good.

9

PRUDENCE

Even if all safety, efficacy and ethical issues were resolved the question would still remain: is it wise to proceed? Human nature and human societies have a way of producing unexpected outcomes from innocuous or well-intentioned innovations. Measures undertaken for apparently good immediate benefits can, over time, cause a shift in social attitudes which is deleterious and may even undermine the altruism which gave the initial motive for the development. Clinical innovation and changes in legislation, if any, ought to be introduced through the democratic process with ultimate decisions being made by parliament, following widespread, detailed and informed public debate. Human genome editing is, arguably, the bioethical equivalent of splitting the atom- an ability which has brought us both nuclear power and nuclear weapons. We ought to proceed with very great care to maximise the benefits for society and guard against its misuse.

While some of what is being proposed in terms of genome editing is in relation to "somatic cells" (i.e. non-reproductive cells, so the changes would only affect the individual in whom the change is made, but would not be passed on to their offspring), others working in the field envisage the editing of genetic sequences in "germ line" cells. These changes would be passed on to subsequent generations; many would argue that the ability to manipulate the genetic sequence of generations yet unborn raises additional questions around the use of such techniques.

It should also be noted that, along to the way to develop these technologies, there were inevitably many failed attempts. Before considering the use of genome editing techniques in humans, it is necessary to be aware that, while the birth of experimentally malformed or mutated mouse pups, kittens or lambs may be seen as morally acceptable, this would not be the case with human babies. How would we care for these children in the future?

It is, of course, not possible to assert exactly where the possibilities opened up by todays technologies will lead in terms of future developments, but various ethical and moral issues are implicit in the technology, which the Church feels it is important to draw to the attention of those involved in the debate-before the technological "horse" bolts from the stable and it is too late to lock the door.

Potential ethical issues may include questions around what is our understanding of what it means to be human: are we more than simply what is determined by our genetic and physical makeup? If intervention in our genetic future is legitimate, what are appropriate changes to that genetic makeup, and what would be improper in terms of our human dignity? Should a distinction be drawn between repairing genetic damage and any potential there might ever be to make genetic "improvements"- and if so, how and where is that distinction drawn? Given that germline genetic editing would be passed on to future generations, what are our duties in respect to the rights and the suffering of future generations and the extent to which humans could intervene in their genome on their behalf?

Clearly, in the drive to develop any new technologies, it is impossible to see in advance all the pitfalls and problems which will almost inevitably arise along with the benefits- there is always a risk involved. However, that does not mean that technology should be developed in and applied in an ethical vacuum. It is important to recognise that, especially in the early stages, it may not be possible to know what all the risks are, until something has gone wrong. History is replete with examples of unexpected or unintended consequences of technologies- especially in the biological context. Side effects of drugs such as Thalidomide, the effect of chlorofluorocarbons (CFCs) on the ozone layer, the accumulation of microbeads or of DDT in the food chain, the introduction and proliferation of alien species such as the rabbit (a benign species in its native Europe) into Australia, where it had no natural predators. It may be possible to take reasonable precautions, but there is no such thing as risk-free technology, any more than there is risk-free life.

It could be argued that to repair defective DNA so that the correct functioning of a gene is restored is a further extension of medicine, fundamentally no different from organ transplants and *in vitro* fertilisation. Somatic genome editing, in which the modification would not be passed on to future generations, could be seen as a special case of medical treatment. However, the familiar issues such as sáfety, unpredictable consequences and consent would assume greater importance because of the nature of genetic disorders. As Christians we would strongly encourage efforts for the alleviation of the suffering- but would argue that it is necessary to always be aware of the potential long term consequences, not only for the individuals, but for society as a whole.

Because somatic genome editing is directed at the body's non-reproductive cells, it should only affect the genetic makeup of that one individual, and not be passed on to any children they may subsequently have. In contrast, any genetic changes in the reproductive cells - germline genome editing - or changes made to the early embryo before the stage of differentiation into reproductive and non-reproductive cells, would affect all future offspring of that person. This raises issues such as what is our duty in respect of future generations, their rights, choices, health, etc., whether it would be possible to ever know enough about the long term effects to judge if it is right to go ahead?

Some may argue that humans have a God-given ability to reason, explore and invent- and thus to NOT use these techniques when the opportunity arises to do so would be wrong. But just because we CAN do something, does that mean that we SHOULD do it?

As Christians, we are called on to take a wider perspective on the issues which affect us as individuals and as society, and to care for the most vulnerable in our society, who may not be able to speak for themselves. As to the way forward, there was considerable disagreement within the reference group. While the majority of the group felt that we should recommend absolute prohibition on the use of genome editing to alter the human germline, some believe that national and international restrictions on the use of genome editing to alter the human germline would be sufficient. While strong guidelines are necessary, we would urge that socially just and safe solutions to the environmental, social and health problems being addressed be sought.

CONCLUSION

Genome editing is seen by many as representing a next step in our ability to analyse and alter the genetics of plants and animals, including ourselves. The notion that knowledge and the choices it offers might be our downfall is as old as the biblical tale of the Garden of Eden. But, in equal measure, history demonstrates the enormous benefits in health and happiness that come with responsible exercise of our intellect and powers of invention. The newest developments in genome editing will demand that we think again about how to balance hope and fear.

GLOSSARY AND LIST OF ABBREVIATIONS

The glossary has been restricted to scientific and medical terms and abbreviations used in the main text whose meanings may not be widely known or understood. Other terms and abbreviations are explained in the text. This glossary is not exhaustive.

Alkaptonuria is a rare inherited genetic disorder in which the body cannot process the amino acidsphenylalanine and tyrosine, which occur in protein.

Allele is a variant form of a given gene. The word "allele" is a short form of allelomorph ("other form") which was used in the early days of genetics to describe variant forms of a gene detected as different phenotypes.

Autologous transplantation (*or* **Autotransplantation)** is the transplantation of organs, tissues, or even particular proteins from one part of the body to another in the same person.

Blastocyst is called the stage of embryonic development that begins about 5 days after fertilisation in humans. The blastocyst consists of the inner cell mass, which will go on to form the embryo, and the trophoblast, which will form the placenta.

CCR5 (C-C chemokine receptor type 5, also known as CCR5 or CD19) is a protein on the surface of white blood cells that is involved in the immune system as it acts as a receptor for chemokines (signaling proteins secreted by cells). This is the process by which T cells are attracted to specific tissue and organ targets. Many forms of HIV, the virus that causes AIDS, initially use CCR5 to enter and infect host cells.

Chemokines *see* CCR5

Cloning means making a genetic copy (a clone) of another individual.

CRISPr/Cas9 is a recently developed method of genome editing.

CVS (Chorionic Villus Sampling) is a technique for obtaining foetal genetic material for use in prenatal diagnosis. A small sample of tissue is taken from that part of the placenta derived from the trophoblast, which shares the foetus' rather than the mother's genotype.

Cystic fibrosis (CF) is a genetic disorder that affects mostly the lungs, but also the pancreas, liver, kidneys, and intestine.

DNA (deoxyribonucleic acid) is a thread-like chain of nucleotides carrying the genetic instructions used in the growth, development, functioning and reproduction of all known living organisms and many viruses. DNA and ribonucleic acid (RNA) are nucleic acids; alongside proteins, lipids and complex carbohydrates (polysaccharides), they are one of the four major types of macromolecules that are essential for all known forms of life.

Embryo is a human individual in the first eight weeks following fertilisation. The term "pre-embryo" is sometimes used to refer to an embryo in the early stages of development before implantation in the uterus, but this usage is controversial and not universally adopted.

Epithelial cells (*or* **Epithelium**) are one of the four basic types of animal tissue, along with connective tissue, muscle tissue and nervous tissue. Epithelial tissues line the outer surfaces of organs and blood vessels throughout the body, as well as the inner surfaces of cavities in many internal organs. An example is the epidermis, the outermost layer of the skin.

Foetus is a developing human individual between nine weeks post-fertilisation and birth.

Gamete is a sex cell (egg or sperm).

Gene is a sequence of DNA or RNA which codes for a molecule that has a function. During gene expression, the DNA is first copied into RNA. The RNA can be directly functional or be the intermediate template for a protein that performs a function. The transmission of genes to an organism's offspring is the basis of the inheritance of phenotypic traits. These genes make up different DNA sequences called genotypes.

Genome editing (*or* **genome engineering**) is a type of genetic engineering in which DNA is inserted, deleted, modified or replaced in the genome of a living organism.

Genotype is the sum total of an individual's genetic characteristics; see also Phenotype.

Germline in a multicellular organism is the population of its bodily cells that are so differentiated or segregated that in the usual processes of reproduction they may pass on their genetic material to the progeny. If the germline is modified, the modification is passed to the offspring.

HFEA (Human Fertilisation and Embryology Authority) is an executive non-departmental public body of the Department of Health in the United Kingdom. It is a statutory body that regulates and inspects all clinics in the United Kingdom providing in vitro fertilisation (IVF), artificial insemination and the storage of human eggs, sperm or embryos. It also regulates human embryo research.

HIV (Human Immunodeficiency Virus) is a virus that causes HIV infection and over time acquired immunodeficiency syndrome (AIDS). AIDS is a condition in humans in which progressive failure of the immune system allows life-threatening opportunistic infections and cancers to thrive.

IVF (In Vitro Fertilization) is a reproductive technology in which egg and sperm cells are mixed together in the laboratory ("in glass", or *in vitro*). The aim is for sperm to fertilise the eggs, which can then be implanted in the intending mother's womb. IVF is sometimes described as "homologous" if the gametes are derived from the partners of a couple who intend to be the child's social parents, and "heterologous" if sperm, eggs or embryos are derived from donors other than the social parent(s).

Mitochondria (*sing* Mitochondrion) are organelles (sub-cellular structures) found within animal and plant cells, concerned with energy metabolism.

Monogenic disorders (*or* **single gene disorder**) is the result of a single mutated gene.

Mutagenesis is a process by which the genetic information of an organism is changed, resulting in a mutation. It may occur spontaneously in nature, or as a result of exposure to mutagens. It can also be achieved experimentally using laboratory procedures. In nature mutagenesis can lead to cancer and various heritable diseases, but it is also a driving force of evolution.

Nucleotides are organic molecules that serve as the monomer units for forming the nucleic acid polymers deoxyribonucleic acid (DNA) and ribonucleic acid (RNA), both of which are essential biomolecules in all life-forms on Earth. Nucleotides are the building blocks of nucleic acids.

PGD (Pre-implantation Genetic Diagnosis) is the technical term of testing an IVF embryo genetically for particular characteristics (such as sex, or particular inherited disease markers) before it is implanted in the womb.

Phenotype is the composite of an organism's observable characteristics or traits, such as its morphology, development, biochemical or physiological properties, behavior, and products of behavior. A phenotype results from the expression of an organism's genetic code, its genotype, as well as the influence of environmental factors and the interactions between the two. When two or more clearly different phenotypes exist in the same population of a species, the species is called polymorphic.

Pre-embryo *see* Embryo.

PUBS (Percutaneous Umbilical cord Blood Sampling) also called cordocentesis, fetal blood sampling, or umbilical vein sampling is a diagnostic genetic test that examines blood from the fetal umbilical cord to detect fetal abnormalities.

Reproductive cloning *see* Cloning.

RNA *see* DNA

SCNT (Somatic Cell Nuclear Transfer) is a laboratory strategy for creating a viable embryo from a body cell and an egg cell. The technique consists of taking an enucleated oocyte (egg cell) and implanting a donor nucleus from a somatic (body) cell. It is used in both therapeutic and reproductive cloning.

Stem cell is a relatively unspecialised cell with the potential to give rise to one or more specialised cell types. Stem cells may be *monopotent* (able to give rise to only one cell type), *multipotent* (able to give rise to a number of cell types), *pluripotent* (able to give rise to all the cell types found in the body) or *totipotent* (able to give rise to all the cell types in the body plus those in the placenta). Stem cells derived from human embryos (human embryonic stem cells, or hESCs) are pluripotent. In recent years it has become possible to turn cells taken from the adult body back into pluripotent stem cells with similar properties to hESCs: these are known as induced pluripotent stem cells or iPSCs.

T cells (*or* **T lymphocytes**) are a type of lymphocyte, a subtype of white blood cell) that play a central role in cell-mediated immunity.

Trophoblast *see* Blastocyst

Zygote is the single-celled stage of embryonic development immediately following fertilisation.

Glossary sources:

- Wikipedia
- *Before I formed you in the womb*, A Guide to the Ethics of Reproductive Medicine from the Council of the Community of Protestant Churches in Europe (CPCE). This book can be downloaded or purchased at www.cpce-repro-ethics.eu

CEC Thematic Reference Group on Bioethics: Membership

Rev. Dr Jean Boboc	Romanian Orthodox Church
Dr Andrea Dörries	Evangelical Church in Germany
Dr Konstantinos Kornarakis	Church of Greece
Bishop Makarios of Christopolis	Ecumenical Patriarchate
Rev. Dr Brendan McCarthy	Church of England
Dr Murdo Macdonald	Church of Scotland
Prof. Ulrik Becker Nissen (Group Chair)	Evangelical Lutheran Church in Denmark
Rev. Dr Miriam Prasilova	Silesian Evangelical Church A.C. in the Czech Republic

Dr Meego Remmel	Estonian Council of Churches
Prof. Luca Savarino	Waldensian Church in Italy
Rev. Dr Miriam Szokeova	Silesian Evangelical Church A.C. in the Czech Republic
Dr Julija Vidovic	Serbian Orthodox Church

Staff members:

Rev. Sören Lenz

Ms Barbara Weber

Made in the USA
Monee, IL
04 November 2026

67379277R00049